Carla Andressa Almeida Farias

Beer production: stages and field application

Carla Andressa Almeida Farias

Beer production: stages and field application

A brief history of the brewing routine in an industry

ScienciaScripts

Cover image: www.ingimage.com

This book is a translation from the original published under ISBN 978-613-9-68504-2.

Publisher:
Sciencia Scripts
is a trademark of
Dodo Books Indian Ocean Ltd. and OmniScriptum S.R.L publishing group

120 High Road, East Finchley, London, N2 9ED, United Kingdom
Str. Armeneasca 28/1, office 1, Chisinau MD-2012, Republic of Moldova, Europe
Printed at: see last page
ISBN: 978-620-8-21853-9

CONTENTS

1 INTRODUCTION

The English word *beer*, translated as cerveja in Portuguese, comes from the Latin *bibere*, which means drink. Among the oldest known drinks is beer.

Most popular among civilizations located in climates not conducive to growing grapes, beer began to be produced around 6000 years ago (REINOLD, 1997). Nationally, at the beginning of the 19th century, King João VI was responsible for introducing people to the habit of drinking beer, which at the time was imported from European countries (VENTURINI FILHO, 2010).

Beer is defined as a fermented alcoholic drink made from water, barley malt, hops and yeast (SILVA, AUGUSTO E POPPI, 2008). Among the constituents of beer, the main ingredient is malt, which is responsible for the product's sensory characteristics, such as taste and color (MARCONE et al., 2013).

Thus, according to current Brazilian legislation for beverages, through Decree No. 6.871 of June 4, 2009, which regulates Law No. 8918 of July 14, 1994, beer must have some specific characteristics, being classified in terms of color, primitive extract, alcohol content, proportion of barley malt and fermentation (BRASIL, 2009; BRASIL 1994).

The most widely consumed beers in Brazil are lager and ale, with lager being the most popular (COOPER et al., 2016). The factors that vary between the types of beer brewed are the form of fermentation, the ingredients and the quantities added, remembering that in addition to the mandatory items determined by legislation, adjuncts can be added to beer, such as corn, rice, unmalted barley, among others. These adjuncts imply a reduction in costs during brewing (KUNZ, MULLER, MATO-GONZALES AND METHNER, 2012; BOGDAN AND KORDIALIK- BOACKA, 2017).

With this in mind, the aim of this work is to carry out a bibliographical review of the entire brewing process, such as raw materials, classifications and stages used in the production of beer, together with an application of all the knowledge acquired during the degree in a brewery in the state of Rio Grande do Sul, with the aim of improving and getting to know the routine of a company.

2. OBJECTIVES

2.1 General Objectives

The aim of this work was to provide food academics with a way of getting to know and gaining experience of a working environment, as well as putting into practice the knowledge acquired during their degree, thus complementing their knowledge and academic training.

2.2 Specific Objectives

- Quality control;
- Monitoring the company's routine activities;
- Understand the beer production process;
- Monitoring the processing stages through physico-chemical and microbiological analysis.

3. BIBLIOGRAPHIC REVIEW

3.1 Definition and history of beer

By definition, beer is the drink obtained by the alcoholic fermentation of beer wort made from barley malt and drinking water, by the action of yeast, with the addition of hops (BRASIL, 2009).

According to the Purity Law or *Reinheitsgebot*, which originated in Germany in 1516 and was approved by Duke William IV of Bavaria, the only ingredients needed to produce the drink are barley, water, yeast and hops (VENTURINI FILHO, 2010).

Although it is difficult to pinpoint the origins of beer production, the literature suggests that the practice of brewing originated in Mesopotamia, where, just as in Egypt, barley was grown in the wild. With this, it can be said that both civilizations have been producing the alcoholic beverage for over 5000 years (VENTURINI FILHO, 2010; REINOLD, 1997).

The brewing process began with the combination of cereals and yeast, raw materials used in baking, obtained after the fermentation of cakes and bread in jars containing water, called "rustic beer" (VENTURINI FILHO, 2010). This beer played an important role in Egyptian culture, such as in religious rituals, bringing knowledge of this product to other peoples and consequently to the rest of the world (VENTURINI FILHO, 2010).

The Greeks learned the brewing technique from the Egyptians and also used hop. The Romans learned from the Greeks and introduced it to the region of Gaul and Spain, but the use of hops only took place after the 8th century (AQUARONE, 1993). The Chinese were the first people to prepare drinks such as beer, obtained from cereal grains, such as 'Samshu', which was made from rice (AQUARONE, 1993).

Europeans discovered the brewing technique not long before the Christian era, when the first drinks were made from a mixture of corn and honey, giving rise to mead, which had a sour taste. In England, the Celts fermented wheat and honey to make beer in the 6th century (AQUARONE, 1993).

Há is also evidence that the Incas, before the arrival of the Spaniards, used to drink beverages made from grâos, known as corn beer (REINOLD, 1997).

Beer was already known in America before Columbus, who was given a corn-based beer by the Indians. "Real" beer was introduced to North America in the state of Virginia by the English in 1548 (AQUARONE, 1993).

Nationally, beer began to be consumed during the colonial period, imported from European countries, especially at the beginning of the 19th century after the arrival of the Portuguese royal family and King João VI, who was a great fan of the drink (VENTURINI FILHO, 2010).

It was only in 1853 that the first national brewery was opened, the Cervejaria Bohemia, located in the city of Petrópolis (RJ), which officially supplied the drink to royalty (AMBEV, 2011).

3.2 Beer classification

The classification of beer depends on the characteristics of the raw materials used, the conditions used, as well as the sensory characteristics of the beer, so its classification will depend on how much (VENTURINI FILHO, 2010):

I - Original extract

Beers are classified according to their primitive extract, as shown in Table 1.

Table 1. Classification of beer according to its primitive extract.

Type of beer	Primitive extract content (% by weight)
Light beer	5,0 a 10,5
Regular beer	10,5 a 12,0
Extra beer	12,0 a 14,0
Strong beer	Above 14.0

SOURCE: VENTURINI FILHO, 2010

II -Color

Clear beer: beer with a color of less than twenty EBC (European Brewery Convention) units;

Dark beer: beer with a color of twenty or more EBC units;

Colored beer: has a different color based on EBC standards due to the addition of colorants (VENTURINI FILHO, 2010).

III - Alcohol content

Alcohol-free beer: beers with an alcohol content of less than 0.5 percent by volume;

Beer with alcohol: beers with an alcohol content of more than 0.5 percent by volume, and the amount of alcohol must be stated on the product label (VENTURINI FILHO, 2010).

Beers can be classified according to their alcohol content, as shown in Table 2.

Table 2. Classification of beer according to alcohol content

Beer	Alcohol content (%)
Low alcohol content	0,5 a 2,0
Medium alcohol content	2,0 a 4,5
High alcohol content	4,5 a 7,0

SOURCE: VENTURINI FILHO, 2010

IV - Proportion of barley malt

Pure malt beer: contains 100 percent barley malt by weight over the original sugar extract;

Beer: contains 55 or more percent barley malt by weight over the original sugar extract;

Beer named after the predominant vegetable: contains between 20 and 55 percent barley malt by weight over the original sugar extract (VENTURINI FILHO, 2010).

V - Fermentation

Low fermentation: this is when the yeast sediments and settles at the bottom of the fermentation tank in a flocculent manner. Low-fermentation beers are classified as lagers and ferment for 7 to 10 days at a temperature of 7 to 15 °C, maturing at 0 °C for 3 to 5 weeks;

Top fermentation: this is when the yeast appears on the surface of the fermentation tank in a flocculent form. High fermentation beers are classified as Ale, and ferment for 3 to 5 days at a temperature of 18 to 22 °C, maturing at 0 °C for a week (VENTURINI FILHO, 2010).

3.3 Components

According to Article 36 of Decree No. 6.871 of June 4, 2009, which provides for the composition of the drink, the basic ingredients for the production of beer are water, malt, yeast and hops (BRASIL, 2009).

3.3.1 Water

Considered the most important raw material in beer production, water accounts for around 90% of the beverage's composition. For this reason, the physical and chemical characteristics of water are extremely important for obtaining good quality beer (MADRID et al., 1996).

It must be potable, so this is one of the decisive factors when choosing a location for a brewery, as it must have an abundant source of good quality water (VENTURINI FILHO, 2000). In addition, it must be clean so as not to interfere with the sensory aspect of the finished beer and free from microorganisms that are harmful to health (GREJO, 2014).

The brewing industry consumes large volumes of water: on average, 10 liters of water are used for every liter of beer produced (REINOLD, 1997). All water to be used in brewing requires preliminary treatment, whether it comes from artesian wells, rivers, lakes or springs. To treat this water, physico-chemical analyses such as color, turbidity, hardness and pH are carried out to determine the appropriate treatment to be used (REINOLD, 2011). An important factor in assessing the quality of water is the presence of heavy metals, as they catalyze reactions with other constituents of beer, allowing turbidity to occur in the medium (LAUX, 1997).

Water has various dissolved salts, as well as organic matter, and the quantity and quality of these compounds contained in water influence the chemical and enzymatic processes that occur during fermentation, altering the quality of the final product, so it is important that it has an adequate concentration of mineral salts, such as calcium (BARBOSA, 2016). The level of calcium in water is of great importance in the brewing industry because, in addition to acting on the stability and taste of the beer, it stimulates enzymatic action (proteases and amylases) by increasing the content of fermentable carbohydrates and nitrogen compounds in the wort, protects a-amylase from thermal destruction, helps control pH by improving yeast yield and flocculation and reduces the color of the wort (BRODERICK et al., 1977; BARBOSA, 2016). Magnesium serves as a coenzyme in the fermentation process (BRODERICK et al., 1977).

Due to the various salts present in water, different specific types of beer have been developed. Thus, water with a high calcium content is associated with beers characterized by greater bitterness, while water with a lower calcium content is suitable for the production of beers characterized by a sweet taste, while pilsen beer requires water with a low mineral salt content for its production (VENTURINI FILHO, 2010).

Controlling the pH of the water is fundamental when brewing beer, as an alkaline pH can lead to the dissolution of materials in the malt, as well as counteracting the effect of calcium and magnesium. Ideally, the pH of the water should facilitate enzymatic activity, increasing the yield of maltose as well as the alcohol content of the beer. This pH should be around 6.5 and 8.0, the range where malt enzymes work to transform starch into fermentable sugars, but the ideal pH depends on the type of beer to be produced (AMBEV, 2014).

3.3.2 Malt

A necessary ingredient for brewing wort, barley is a grass of the genus Hordeum (BRIGGS, 2001). Its kernels on the cob can be lined up in two or six rows. The two-row cob is the most widely used because it has more developed and denser kernels, thus providing a higher yield (KUNZE, 1996).

Barley is used because it has a high content of starch, protein and important enzymes. The ideal type of barley for malting is one with a high germination power and low protein content, because the lower the protein content, the more soluble the malt will be (BRODERICK et al., 1977).

One way of classifying barley is by its sowing season. Winter barley is sown in winter and summer barley is sown in spring. Winter barley has a higher yield in the field than summer barley (ZSCHOERPER, 2009).

Barley grain is made up of three main parts: the endosperm, which is basically made up of starch; the husk, which is insoluble and protects the grain from harmful atmospheric influences, making it possible to form a filter layer during the separation of the wort from the solid matter; and the embryo which, under the right conditions, germinates, transforming the endosperm and activating enzymes, being of great importance in the production of must (BRODERICK et al., 1977). Due to its composition, barley is the main material for brewing beer.

Malt is obtained through the germination of a particular cereal, barley being the most suitable for this process, as it provides a superior flavor compared to other cereals (REINOLD, 2011). The malting process is divided into three parts: maceration, germination and drying. In maceration, the barley grain is immersed in water, creating a condition for the grain to germinate; in germination, as the name implies, the grain will germinate, forming enzymes and breaking down the grain; in drying, the grain is heated to approximately 70 °C, reducing its humidity. The aim of malting is to obtain enzymes (AQUARONE, LIMA, BORZANI, 1983).

3.3.3 Lùpulo

Responsible for the sensory characteristics of the drink, the lupulus is a 5 to 7 meter high vine that belongs to the family of moraines, known as *Cannabinaceae*. This plant produces female and male flowers. The female flowers are grouped in clusters or umbels which have a vein with several folds, on which are attached bracts and bracteoles. These bracts and bracteoles are pockets in which lupulin granules are stored, which is why these granules are of great interest to the brewing industry. These granules contain aromatic and volatile compounds, which act as natural preservatives. Its bitter resins are what give the product its bitter taste (VENTURINI 2005).

Hops can be sold in the form of dried flowers, powder and extracts. In powder form, they have a higher density and take up less space for storage and transportation (TELES, 2007). It is fundamental in the production of beer, as its components enable sensory characteristics such as bitterness and aroma due to essential oils, minerals and tannins (REINOLD, 1997).

3.3.4 Yeast

Brewer's yeast is used to ferment the wort (BRASIL, 2009). Yeasts are unicellular fungi and are responsible for alcoholic fermentation and CO_2 production. The yeast, temperature, fermentation pH, fermenter model and wort concentration are all factors that influence the taste of the beer during the fermentation stage (CARVALHO, BENTO E SILVA, 2006).

The yeasts most commonly used in the brewing industry are those of the genus *Saccharomyces cerevisiae* for high fermentation, at a temperature of 18 to 22 °C for 3-5 days. This yeast flocculates on the surface of the tank, producing ale. The yeast of the genus *Saccharomyces uvarum* for low fermentation, at a temperature of 7 to 15 °C for 7-10 days, causes the yeast to flocculate at the end of fermentation, settling at the bottom of the tank, producing lager beer (LAUX, 1997).

The metabolic difference between ale and lager strains is the ability of lager to ferment the sugar melibiose, due to the presence of the enzyme a-galactose or melibiase, which transforms melibiose into glucose and galactose. This disaccharide, which has been transformed, together with fructose forms the trisaccharide raffinose. As a result, lager yeast can fully metabolize raffinose, while ale yeast, because it doesn't contain melibiase, metabolizes only part of the raffinose, metabolizing only fructose (REINOLD 1997).

3.3.5 Optional ingredients

In addition to the mandatory ingredients for making the drink, according to the legislation, part of the barley malt can be replaced by brewing adjuncts, however, their use cannot exceed forty-five percent of the original extract (BRASIL, 2009). Adjuncts are brewer's barley and other cereals suitable for human consumption, such as starches and sugars of vegetable origin (BRASIL, 2009).

Venturini's (2005) definition of adjuncts as "malted or non-malted carbohydrate materials of appropriate composition and properties that beneficially complement or supplement barley malt" is also worth mentioning. These adjuncts are used with the intention of reducing production costs and providing peculiar characteristics to the final product.

The main adjuncts used are sugars, syrups and unmalted cereals, such as sucrose, glucose, invert sugar, high-maltose, wheat, rice, corn, oats, among others. The purpose of sugars and syrups is to attenuate specific characteristics of the beer and provide color, flavor and aroma (VARNAM AND SUTHERLAND, 1997).

In Brazil, the most commonly used adjuncts are rice grits and ground, degerminated corn, which produce a sweeter, fuller-bodied beer (REINOLD, 1997).

3.4 Beer processing

The processing of beer can be divided into three phases: the first is the production of the wort, through the processes of milling the malt, mashing, filtering, boiling and cooling; the second consists of the fermentation stage, covering fermentation and maturation; and the third is the finishing phase, through filtering, carbonation, bottling and pasteurizing the beer (AQUARONE, LIMA AND BORZANI, 1983).

3.4.1 Malting

The aim is to reduce the size of the malt kernel uniformly by breaking the kernel husk, exposing the endosperm, so that only the endosperm is crushed and not the husk, thus ensuring total disintegration of the endosperm and promoting better enzymatic action; You also want to obtain a minimum amount of flour from the grain so that an excessive amount of paste doesn't form in the solution (VENTURINI FILHO, 2010).

When the grind has very large particles, this can make it difficult to hydrolyze the starch - hydrolysis is an important process because starch is converted into sugar. However, when the grains are very small, in the form of flour, this can lead to problems when it comes to filtering them afterwards, resulting in the flour being passed on to the other stages of brewing, as well as the formation of paste in the solution. Therefore, for ideal milling, there should be no whole kernels, most of the husks torn lengthways, no endosperm particles adhered to the husks, endosperm reduced to uniformly sized particles and a minimum amount of flour. This can be done using roller mills (AQUARONE, 2001; VENTURINI FILHO, 2010).

3.4.2 Mashing

Its aim is to solubilize the water-soluble substances in the malt and, through the action of enzymes, to solubilize the insoluble substances in the malt, providing gumification and hydrolysis of the starch to sugars. It is a temperature- and time-controlled process, as shown in Figure 1 (VENTURINI FILHO, 2010).

Enzymes act in various biochemical transformations and are greatly influenced by the temperature and pH of the medium. They are unique, acting in their specific ranges, and their parameters are taken into account. The most important enzymes in beer are beta-amylases and alpha-amylases. Beta-amylase has the characteristic of breaking down starch into smaller, fermentable sugars, in this case breaking down starch into maltose, which results in the alcoholic strength of the beer. Alpha-amylase also breaks down starch, but its breakdown results in sugar chains of various sizes, which are not fermentable like dextrins, so its breakdown gives the beer body and sweetness (AQUARONE, 2001).

These are not the only enzymes at work in beer, there are others that provide aspects of foam, aroma and turbidity to beer, as shown in Table 3.

Table 3. Temperature and pH at which the enzymes work

Enzymes	Temperature	pH	Performance
Phytase	30 to 52 °C	5,0 a 5,5	Decreases mash pH
Beta-Glucanase	35 to 45 °C	4,5 a 5,5	Breakdown of glucans
Proteases	45 to 55 °C	5,2 a 5,8	Reduces turbidity and forms foam
Beta-Amylase	60 to 65°C	5,4 a 5,6	Checks alcoholic strength
Alpha-Amylase	70 to 75 °C	5,6 a 5,8	Gives the beer body and sweetness

SOURCE: http://www.cervejacsac.com/mundo-cervejeiro/enzimas/

It is worth noting that the enzymes become inactive above 75 °C.

Figure 1. Temperature variation as a function of time during the mashing process.

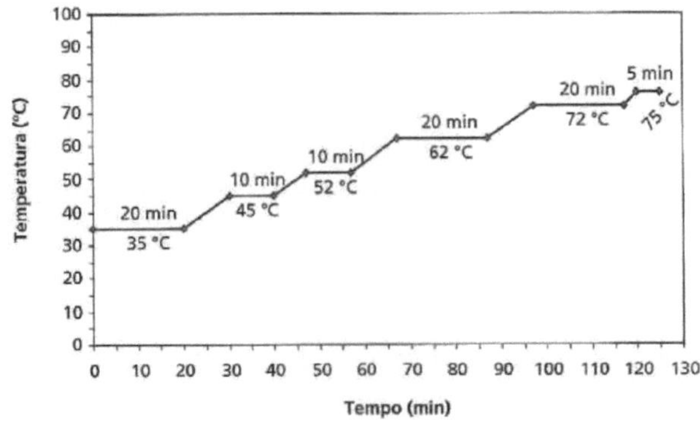

SOURCE: VENTURINI FILHO, 2010

3.4.3 *Wort filtration*

Filtering is carried out in a filtering vat, where the malt husk serves as a filtering layer, this layer is washed with water at 75 °C in order to increase the extraction of sugar present in the husk and increase the yield of the process (VENTURINI FILHO, 2010).

3.4.4 *Boiling*

Boiling takes place with the must filtered and with the addition of hops. This stage aims to inactivate enzymes, sterilize the must, coagulate proteins, extract bitter and aromatic compounds from the hops, form aroma and flavour constituents and evaporate excess water and undesirable aromatic compounds in the product. When boiled, it is necessary to remove the precipitates (trub) present in the must, followed by cooling and aeration (VENTURINI FILHO, 2010).

3.4.5 *Fermentation*

The aim of fermentation is to convert sugars into ethanol and carbon dioxide gas by the action of yeasts under anaerobic conditions. Yeasts produce compounds that give beer its aroma and flavor. Fermentation can take place with both top-fermenting and bottom-fermenting yeasts (VENTURINI FILHO, 2010).

3.4.6 Maturation

It is a process in which the beer is stored in tanks at low temperatures, developing reactions that give the beer sensory characteristics such as aroma and flavor. It also clarifies the beer due to the precipitation of yeast, proteins and insoluble solids. Diacetyl can form during the fermentation of the wort and can be eliminated during maturation. Its presence is considered a defect depending on the amount it is found, as it can give the beer a rancid butter taste (VENTURINI FILHO, 2010).

3.4.7 Filtration

The filtration process takes place to clarify the beer, i.e. it is carried out to remove suspended material, as well as residual yeasts that promote beer turbidity (VENTURINI FILHO, 2010).

3.4.8 Carbonation

Carbon dioxide (CO_2) is responsible for the effervescence and sensation of acidity left in the mouth. For this reason, the injection of CO_2 during production is extremely important in order to guarantee consumers a quality product (VENTURINI FILHO, 2010).

3.4.9 Packaging

Filling is the bottling, canning or kegging of the product, in this case the filling of beer. It is an operation carried out in equipment known as a filler when it comes to bottles and cans or in kegging machines when it comes to barrels (VENTURINI FILHO, 2010).

3.4.10 Pasteurization

Pasteurization is the destruction of microorganisms in aqueous solutions by heat at a temperature of 60 °C for one minute. Beer is a product favorable to microbiological growth, so this growth can lead to undesirable characteristics in beer, such as turbidity and sensory changes, and pasteurization is used to try to avoid these problems (VENTURINI FILHO, 2010).

4. ACTIVITIES CARRIED OUT

The routine of a quality control (QC) laboratory in an industry requires a lot of attention, agility, responsibility and care when carrying out activities, because it is QC that determines whether the product being processed is of good quality, so that it can be put on sale.

The company where the internship took place was divided into three sectors: (I) Manufacturing Sector, (II) Filling Sector and (III) Microbiology Sector.

4.1 - (I) Manufacturing Sector

It consists of brewing activities, checking the fermentation tanks, filtering the beer and releasing the pressure tanks.

4.1.1 Brewing

Brewing was separated into five main stages: milling, mashing, filtering, boiling and cold wort. Among these stages, it was necessary to carry out physico-chemical analyses (described below) to determine whether the beer being processed complied with the specific parameters for each brewing stage.

a) Process water and brewing water

The water was used to start the wort boiling process. The pH of the water was then analyzed and it was boiled at 78 °C.

- pH analysis

pH was measured as the potential difference between an electrode and the liquid under analysis, making it possible to selectively measure the concentration of a particular ion. The pH analyses were applied to all types of wort and beer.

The working pH range for process and brewing water was 5.00 to 5.50, and this was corrected by adding water (increasing the pH) or lactic acid (decreasing the pH). The pH range used was aimed at good enzymatic activity, since too high a pH in the washing water could cause unwanted extraction of tannins from the malt cake. In the case of the mashing water, a pH outside the ideal range was not too serious, as the malt grains helped to balance the mash pH.

b) Grinding

At this stage, the proportions of husk, grain and flour resulting from the milling of the malt were analyzed. In this way, in a container containing the milled malt, it was sieved through three sieves of different sizes, which provided quantities of husk, grits and flour, giving a percentage view of the grain size of the malt. Adjustments to the roller mill were made based on the particle size analysis carried out in the laboratory.

The ideal grind was very relative, the most common being the total or partial removal of the husk, freeing the endosperm for subsequent mashing. The grind must be suited to the filter media used during manufacture. The company had two brewing facilities, the micro brewery with a capacity of 70 hectoliters (hl) and the large brewery with a capacity of 220 hl. In micro brewing, filtering was done using a tank with a false bottom, in which case it was ideal to use malt with all or part of the husk removed, since the malt itself served as a filtering medium, giving a higher yield at the end. However, in large-scale production, a method of forced filtration with pressure and a higher final yield (plate filter) was used. In this case, the malt had to be finely ground because it didn't need the husk as a filtering medium.

Very fine grinding could facilitate the extraction of unwanted substances, such as the tannins in the bark that give the beer its astringency, but temperatures below 80 °C make this type of extraction difficult. And coarse grinding can cause a loss of efficiency, affecting the final density of the beer.

c) Mashing

In the mashing stage, the ground malt was mixed with the process water. This stage came from a heating ramp, a temperature and time-controlled ramp throughout which the temperature of the wort was raised to approximately 76 °C.

At this stage, initial and final pH analyses of must, extract and saccharification were carried out.

- *Extract Analysis*

Extract analysis was carried out to determine how much sugar was in the medium. The sample received was filtered through filter paper and soil and read on an analog refractometer, with the result obtained in °Brix.

- *Saccharification Analysis*

The iodine test, also known as saccharification, consisted of comparing the color of pure iodine with the mixture of iodine and must, in order to determine whether the must had starch in its composition. If the color of the mixture turned blue, it meant that the must still had starch molecules in it, requiring a longer mashing time to break down the starch molecules completely.

As the raw materials used to make beer contain a large amount of starch, this starch must be converted into fermentable sugar. In this way, the enzymes present in the malt have the task of transforming the starch into sugar and making the proteins soluble.

d) Filtration

After all the starch had been converted into sugar in the mashing stage, the wort was filtered through a filter press, which retained the malt husks and the pomace from the liquid part. At this stage, the filtered wort was analyzed for turbidity and extract.

- *Turbidity analysis*

Turbidity was measured in musts, beers that were still being processed and finished products, using a spectrophotometer reading at a wavelength of 700 nm to obtain the turbidity value, measured in Nephelometric Turbidity Units

(NTU).

e) Boiling the wort

Once the wort had been filtered, it was transferred to the boiling vat, where the temperature was raised to approximately 100 °C. This boiling took place in two stages, one with the original wort and the other, when necessary, with the addition of maltose. Maltose was used to increase the wort extract, which was economically viable as it reduced the wort boiling time; in beers where maltose was not added, if the extract was low, it was necessary to boil the wort longer, or increase the evaporation rate, as well as add a specific volume of maltose or wort mixture from later brewing. At this stage, hops were added, responsible for the bitterness when added at the beginning of the boil and for the aroma when added at the end, as well as brewing preservatives.

This stage also saw the inactivation of enzymes and the coagulation of undesirable proteins, resins and tannins into lumps, known as "trubs".

The analyses carried out at the start of the boil were: pH, extract and saccharification, while those carried out at the end of the boil were: pH, extract without and with maltose, color, Indicator Time Test (ITT), redox value, saccharification and decantation test (trubs).

- *Color Analysis*

The color was measured using a spectrophotometer and read at a wavelength of 430 nm, resulting in a color value according to the EBC scale.

- *Indicator Time Test (ITT) analysis*

This analysis determined the stability of the beer's flavor during handling and storage, through the reduction potential and presence of antioxidants in the beer.

The analysis was carried out in a spectrophotometer reading at a wavelength of 540 nm. First, the absorbance of the beer was read and the value obtained was added to the correlation factor of the ITT solution, then 2,6-dichloro phenol indophenol (DCI) was added to the beer, and with the help of a stopwatch the time in seconds that the reaction took to reach the value of the sum of the beer and the ITT solution factor was determined. The answer was given in seconds and the ideal time depended on the characteristics of the beer.

- *Redox Value Analysis*

Strongly reducing substances are easier to oxidize and have a lower activation energy. In samples with strongly reducing substances, the agent will be reduced and consumed quickly, while in samples with weakly reducing substances, the agent will decrease more slowly.

In this way, the redox value was determined using a spectrophotometer and read at a pre-established wavelength based on the pH of the must. An ITT solution was added to the must sample and the wavelength was then read at 3 and 8 minutes. Equations 1, 2 and 3 were used to determine the result. Where X and Y are the values for 3 and 8 minutes and Fc is the value for the cofactor.

$$Z1 = \frac{1}{X} \quad (1) \quad ; \quad Z2 = \frac{1}{Y} \quad (2)$$

$$W = (Z1 - Z2)\,.\,Fc \quad (3)$$

- *Trubs analysis*

It measures the amount of coagulated substances present in the must. In this way, the collected must was poured into a decanting cone with a maximum capacity of 1000 mL and left to stand for 1 hour, in order to determine the value of trubs per mL.

Between boiling and cooling there was the whirpool - responsible for decanting most of the residual protein and hop from the boil. Its main purpose was to clarify the wort before cooling.

f) Cold wort

This stage was based on cooling the must using a plate heat exchanger, with the aim of lowering the boiling temperature to a suitable temperature (usually 10 °C) for subsequent fermentation. Before carrying out the analyses for cold must, a sample of the must was taken for microbiological analysis.

The analyses carried out for cold must were pH, bitterness, calcium, color, extract, Indicator Time Test (ITT), redox value, temperature and trubs.

The cooling temperature of the must was checked on a digital panel.

- *Bitterness analysis*

Bitterness analysis was applied to samples of hopped wort, beers in process and the finished product. The substances that give beer its bitterness were extracted from the beer in an acidic medium, in which case we used iso-octane as a reagent, which gave the solution a phase separation when it was subjected to rest. The bitterness in the beer was measured using a spectrophotometer, reading at a wavelength of 275 nm, resulting in a bitterness value of IBU (International Bitterness Unit).

- *Calcium analysis*

This analysis was carried out on cold must and the final product. It was a titrimetric technique and consisted of a solution in which the must was diluted in distilled water, followed by the addition of sodium hydroxide, followed by a hydroxynaphthol blue indicator with EDTA titrant. To determine the amount of calcium in the must, the calculation was carried out as follows: multiply the amount spent on the titration by 40.00 (molar mass of calcium) and the EDTA correlation factor.

4.1.2 Fermentation tanks

After the beer was brewed, the wort was added to the yeast and stored in large fermentation tanks. The fermentation tanks were basically controlled in two stages: the fermentation and maturation of the beer, and the subsequent release of the tank for filtration.

During fermentation, the yeast transformed the sugar in the wort into alcohol and carbon dioxide. This stage took place at a controlled temperature depending on the type of beer to be fermented. During fermentation, the beer was not yet ready for consumption. It is worth noting that this stage was the most important for the taste of the beer, as the yeast produced substances responsible for the aroma and flavor of the beer in low quantities.

The fermentation tanks were checked daily for pH, apparent extract, vicinal dicetones (VDK) and temperature and pressure in the tanks.

For the cold start and the beginning of the ripening stage, the apparent extract should give a similar value for three consecutive days, and its VDK should be below 0.15, i.e. it should not taste like rancid butter.

- *Apparent extract analysis*

The apparent extract was determined with the help of a saccharimeter, which had the function of a densimeter. This analysis was carried out by immersing the saccharimeter in the must, which determined the value in degrees Platò (°P), together with the temperature at which the must was present.

- *Analysis of VDK (Vicinal Dicetones)*

This analysis was carried out for fermenting beers, which, by steam distilling the beer, would obtain diacetyl and 2,3-pentadione, adding o-phenylenediamine to measure the absorbance at a wavelength of 335 nm.

Diacetyl is a substance that is formed extracellularly by the spontaneous oxidative decarboxylation of α-acetohydroxybutyrate and a- acetolactate. These acids leave the cell during fermentation and are intermediate products in the synthesis of valine and leucine. The compound can give the drink an odor of rancid butter, honey or caramel. The perception limit for diacetyl is around 0.15 mg L^{-1} (REINOLD, 1997).

VDK analyses were carried out when the apparent extract was around 5.00 and the beer wort would only pass from the fermentation stage to the maturation stage when its VDK value was below 0.15.

In the maturation stage, or cold opening, the temperature of the tanks was lowered continuously until it reached 0 °C, where the beer would be released to start filtering. In this way, the beer would be cooled and most of the yeast would be separated by decanting, as it would stop working in the beer. At this stage, the beer's flavor and aroma were enhanced by the formation of esters. During maturation, the temperature and pressure of the tanks were controlled.

When the temperature reached 0 °C, the beer was released for filtration, and tests were carried out for pH, bitterness, color, calcium, VDK, turbidity, apparent extract, original extract, real extract, alcohol (% v/v), carbon dioxide (CO_2) and dissolved oxygen (DO).

- *Analysis of original extract and real extract*

The original extract was carried out on beers to determine the wort extract from the fermented beer. Determining this extract was one of the ways of classifying beer.

The real extract represented all the solids that were part of the beer's composition, as well as determining the amount of sugar remaining in the beer after fermentation.

The analysis of original extract, real extract and alcohol (% v/v) was carried out on a device called Alcolyzer Beer, for which a quantity of sample was injected, followed by its reading on the panel.

4.1.3 Beer filtration

When the beer had finished maturing, it was filtered to remove suspended particles, making the product cleaner. This filtration took place with diatomaceous earth, which is considered a filtering material that is designed to sediment the particles formed during maturation by gravity.

The analyses carried out for filtration were color, turbidity and apparent extract, which were carried out every 30 minutes. When the filtration process was complete, the beer was transported to pressure tanks so that it could be bottled.

4.1.4 Pressure tanks

The purpose of the pressure tanks was to provide additional carbonation to the beer when fermentation did not meet the necessary CO_2 demand described in the specifications, with the can being 0.51 to 0.53 and for bottles 0.53 to 0.58. The temperature of the tanks had to be between 0 and 1 °C to preserve the product.

The analyses carried out to release the tanks for subsequent filling were: bitterness, sulphur dioxide, Indicator Time Test (ITT), alcohol content, carbon dioxide, color, dissolved oxygen, pH, apparent, original and real extract. A sensory analysis was also carried out in order to observe characteristics relating to the product's appearance, taste, aroma, touch and drinkability through a sensory analysis.

4.2 (II) - Filling Sector

During the training in the filling sector, activities were carried out regarding the bottle washer, fillers, pasteurizer and pasteurization unit, visual inspection of the product, packaging volume and analysis of the finished product.

4.2.1 Bottle washers

The washers were used to wash the returnable bottles from outside. The purpose of the washers was to remove all dirt from the bottles for later use.

The bottle washer had four baths containing caustic soda (NaOH). These washers were controlled in terms of temperature and the concentration of NaOH present in each bath. The temperature was monitored by a digital thermometer and the concentration of NaOH was measured by titration, where NaOH was added together with a phenolphthalein indicator, the titrant being a solution of hydrochloric acid. In this way, the amount spent on the titration referred to the NaOH concentration of the baths.

Table 4 shows the concentrations of caustic soda and the temperature that the bottle washer baths should be at. Table 4. Concentration of caustic soda and temperature of the washer baths.

Baths	NaOH concentration (%)	Temperature °C
1	1,5 a 2,5	60 a 65
2	1,8 a 2,5	75 a 80
3	Less than 1.5	65 a 75
4	Less than 0.5	Less than 60

SOURCE: BREWING COMPANY

Inside the washer, there was a label extraction system for subsequent internal washing of the bottles.

the presence of an operator who visually inspected the bottles for the presence of labels after extraction, dirty, cracked and broken bottles.

- *Dirt test*

As for dirt, methylene blue was added to the bottles and a light panel made it possible to see dirt on the bottles due to the dark spots present.

- *Test for the presence of caustic soda*

To determine the presence of caustic soda (NaOH), the indicator phenolphthalein was used, which, in the presence of NaOH, turned pink. In this way, the bottles were washed with distilled water, and if the water turned pink when the indicator was added, it was because the bottles had not been washed properly and contained NaOH.

When the bottles didn't pass the dirt and caustic soda test, a large number of bottles were taken and put through the washer again.

4.2.2 Fillers

This stage was used to fill the cans and bottles with beer from the pressure tanks. However, before starting the filling process, it was necessary to carry out the "filler turn", where a sample was taken to carry out analyses of carbon dioxide (CO_2), total air, color, foam stability, "go no go" (cap sealing), Indicator Time Test (ITT), pH, turbidity, apparent extract and dissolved oxygen, in order to release the beer for filling.

At the filler, the cans and bottles were sealed after the beer was added. The purpose of this sealing was to check that the cans were properly sealed, thus avoiding possible leaks. This check consisted of measuring the height, thickness, lid hook and body hook of the cans. This was done before filling began, because if the cans were not measured within the permitted parameters, the filler had to be adjusted. Bottles were capped "go no go", also known as "pass-no-pass", and with the help of a fixed diameter gauge, the seals on the bottles were observed.

- *CO analysis$_2$ and total air*

Its determination was important because if the beer had a low concentration of CO_2 it would become "shocky", losing important characteristics of the beer, and if the concentration was high, it could lead to the packaging breaking.

The Zahm-Hartung CO_2 meter was connected to a container containing caustic soda and had a device capable of piercing the tops of bottles and the bottoms of cans. In this way, two cans and bottles were taken and placed in a 55 °C bath to leave the packages at room temperature, then one of the packages was analyzed in the Zahm-Hartung by stirring to find out the maximum pressure reached by the sample, the other package was analyzed for temperature using a digital thermometer, through a relationship between the pressure values inside the packages and the temperature of the beer, the CO_2 content was given from a table. If the CO_2 value found was outside the standards, the pressure tank had to be recharged. The amount of CO_2 for cans and bottles was given as a percentage (%p/p) as shown in Table 5.

Table 5. Permissible percentage of CO_2 for cans and bottles.

Packaging	Amount of CO_2 (%)
Can	0,50 a 0,53
Bottle	0,53 a 0,58

SOURCE: BREWING COMPANY

The same packaging used to analyze CO_2 was used to determine total air, where the packaging was depressurized in the Zahm-Hartung and the amount of air inside the packaging was observed due to the formation of air bubbles in the device's burette. The amount of total air for cans and bottles was given in mL/package as shown in Table 6.

Table 6. Allowable percentage of total air for cans and bottles.

Packaging	Total air quantity
Can	Less than 0.5 mL
Bottle	Less than 0.8 mL

SOURCE: BREWING COMPANY

4.2.3 Pasteurizer

After the cans and bottles had passed through the fillers, they went to the pasteurizer. The pasteurizer consists of a tunnel containing nine internal baths, which are monitored for pH and temperature. In terms of temperature, the pasteurizers had cooling and heating zones.

The purpose of the pasteurizer was to control the microbiological stability of the beer by raising the temperature, thus inhibiting the proliferation of microorganisms.

One of the reasons for controlling the pH of the pasteurizing baths was to guarantee the integrity of the packaging, because when the pH is outside the established standards, it can cause the cans to turn yellow.

The pasteurizing baths varied in temperature from 25 to 62 °C and were measured using a digital thermometer. The most important baths were 4, 5 and 6, where the beer was pasteurized; the others were used to lower the temperature of the product using a lower temperature and not suffering thermal shock in the packaging.

When the pH was outside the permitted standards, phosphoric acid was added for high pH and sodium hydroxide for low pH. When the temperature was outside the standards, the machine operator was told to readjust it.

Table 7 shows the pH range that pasteurizing baths should be in for cans and bottles.

Table 7. pH values of pasteurizing baths

Packaging	pH range
Bottles	7,0 a 8,0
Cans	6,5 a 7,5

SOURCE: BREWING COMPANY

4.2.4 Pasteurization Unit

The pasteurization unit (UP) was a process used to control the time and temperature at which cans and bottles passed through the pasteurizer, in order to obtain a biologically stable beer as the final product. This control was carried out with the help of a device known as a Redpost and was determined constantly during the filling of the beer.

Biological death occurred when the beer was heated to 60 °C for one minute. The Redpost device determined the time the product was in the pasteurizer, the maximum temperature reached, the time the product remained at this temperature and the temperature at which it left the pasteurizer.

The UP values vary from beer to beer. In this case, when the UP value was below or above what was allowed, the filling process for the batch produced was stopped and finished products were randomly removed for further analysis. When the UP values were low, it meant that the temperature reached was not enough to eliminate the micro-organisms, when it was above, the temperature reached exceeded too much, causing the product to cook, consequently changing the product sensorially.

When the UP values were within the established standard, unpasteurized and pasteurized samples were taken for microbiological comparison.

4.2.5 Visual inspection

Inspections of cans and bottles were carried out after the product had passed through the pasteurizer. The inspections differ for cans and bottles.

The cans were inspected to see if there were any strange, leaking, scratched or dented cans, if the cans contained words such as "contains gluten and barley", if all the printing was present, such as the batch, expiry date and if the printing was not blurred, illegible or decentralized. The bale with the cans was also inspected, checking for defects in the plastic, whether the plastic was torn and whether it lacked strength.

The bottles were inspected for the number of "c" bottles, whether they were leaking or cracked, and the labels were analyzed for lithography and wording, whether they were glued in line, without excess glue, torn, detached or wrinkled, and whether there was no label. The caps were inspected for wrinkles, scratches or defective wording, and the caps were also analyzed for the sealing of the bottles. A box of beer was also inspected.

When the cans and bottles showed problems with the visual inspection, they were removed and discarded.

4.2.6 Packaging volume

The volume of the can and bottle packaging was checked, as both had to comply with INMETRO's Volume Control, so the volume was checked using a sensor after the beer left the pasteurizer and randomly on a digital scale.

In order to confirm the volume of the packaging, random samples were taken and the finished product was initially weighed, followed by weighing the product with the addition of water up to the top of the packaging, then weighing the packaging containing only water and the empty packaging, together with the density of the product, it was possible to find out the final volume of the finished product.

INMETRO considers a difference of up to 3% between the volume shown on the label and the actual volume to be acceptable. The company deducts 2.5% of this volume to be within the standard (INMETRO, 2009).

4.2.7 Finished Product Analysis

After the beer had gone through all the brewing and bottling stages, analyses were carried out to ensure that the quality of the product was within the established standards. Analyses were carried out on alcohol (% v/v), alcohol (% w/w), apparent, primitive and real extract, ITT, VDK, color, pH, bitterness, turbidity, foam, density, apparent attenuation, sulfur dioxide, determination of total air and carbon dioxide, "forcing test" and polyphenols, as well as sensory analysis of the finished product.

4.3 (III) - Microbiology Sector

The function of the microbiology department is to check that all products are free from microbiological contamination, in order to guarantee product quality and food safety for consumers. For this reason, the temperature of the incubators was checked daily, the fermenters and ripeners were collected, the cold must and the cleaning water were collected. They also organized and cleaned the materials in use and prepared the culture media.

In the microbiology department, colonies were counted, followed by analysis using gram staining to determine the molds, yeasts and bacteria present in the samples.

4.3.1 Gram stain

Gram staining determined whether the bacteria grown in the culture medium were gram-positive or gram-negative. The gram-positive bacteria had a thicker cell wall and were more resistant to the staining of the reagents, while the gram-negative bacteria had a thinner cell wall and were both spoilers and contaminants.

In order to determine which type of bacteria had developed in the culture medium, the bacteria's color was analyzed. If the bacteria colored purple, they were classified as gram-positive; if they colored pink, they were gram-negative. These analyses were carried out using a microscope.

4.3.2 Dead and suspended cell count

To carry out the count, the sample under analysis was homogenized in a porcelain dish with the methylene blue reagent. The purpose of this reagent was to incorporate dead yeasts, giving them a blue color.

From this homogenized sample, it was transferred to a Neubauer plate where it would be observed under a microscope.

This analysis was carried out for yeasts after CIP (cleaning solutions), when the yeast was changed in the tanks and before each production. In the fermenter, it was analyzed after 24 hours of fermentation and in the maturer it was done immediately after the complete release of the maturation tank for subsequent filtration.

4.3.3 Mezzanine

The mezzanine room was nothing more than a storeroom where beers from all the batches produced in the company were stored. The purpose of this room was to store samples of the finished product until the end of its shelf life, in case it was necessary to carry out any analysis while the beer was still on its shelf life.

Every month, a sensory analysis was carried out on all the batches of beer brewed, to check that there were no sensory changes in the product.

4.4 Other activities carried out during the internship period

In addition to the activities typical of each laboratory, cleaning solutions and water analyses were also carried out.

4.4.1 Analysis of cleaning solutions (CIP)

The solutions used to clean equipment, tanks, filling machines and barrels were CIP systems. In these systems, the cleaning and rinsing solutions were circulated until the containers were completely sanitized.

The CIP solutions consisted of sodium hydroxide, chlorine, nitric acid and peracetic acid. They were used in a controlled manner, with titration analyses being carried out and each solution having its own control range with the permitted solution limits, as shown in Table 8.

Table 8. Range used for CIP solution concentrations

Solutions	Manufacturing	Packaging
Nitric Acid	1,5 a 2 %	1 a 1,5 %
Peracetic acid	0,3 a 0,4 %	0,3 a 0,4 %
Chlorine	Less than 30 ppm	80 to 150 ppm
Sodium hydroxide	3 a 3,5 %	2,5 a 3 %

SOURCE: BREWING COMPANY

- *Nitric acid analysis*

To determine the concentration of nitric acid, a sample was added to an erlenmeyer flask along with the methyl orange indicator and titrated with a sodium hydroxide solution.

- *Peracetic acid analysis*

To determine it, the sample was added to a flask along with sulphuric acid. It was titrated with a solution of potassium permanganate until a pink color was present; then potassium iodide was added and titrated with sodium thiosulphate. To determine the concentration of peracetic acid, the volumes used in both titrations were added together.

- *Chlorine analysis*

For its determination, the sample was added to an erlenmeyer flask along with acetic acid, potassium iodate and starch, and titrated with a solution of sodium thiosulphate to determine the chlorine concentration.

- *Sodium hydroxide analysis*

To determine it, a sample was added to an erlenmeyer flask along with the phenolphthalein indicator and titrated with a solution of hydrochloric acid. The amount spent on the titration was the percentage spent on sodium hydroxide.

When the value for CIP solutions fell outside the permitted standards, the operator was notified so that it could be corrected, after which the sample was analyzed again to see if the correction had been made correctly.

4.4.2 Water analysis

Physico-chemical analyses were carried out on all the company's water units (boiler water, cooling towers, supply water and waste water used to clean the filters in fermentation tanks in cleaning solutions).

The analyses carried out were:

- *pH, conductivity and total dissolved solids*

It was determined using a device capable of measuring all the parameters, and the reading was taken with the help of an electrode. The electrode was dipped into the sample, and the value corresponding to the analysis was given by a digital panel.

- *Alkalinity*

The titration method was used to analyze alkalinity. A sample was added to an erlenmeyer flask along with the methyl orange indicator, after which the solution was titrated with sulfuric acid. When the color of the solution turned orange, the amount spent on the titration to obtain the alkalinity value was noted.

- *Chlorides*

For the analysis of chlorides, the titration method was also used, where a sample was added to an erlenmeyer flask along with potassium dichromate used as an indicator, after which the solution was titrated with silver nitrate. When the color of the solution turned brick red, the amount spent on the titration to obtain the chloride value was noted.

- *Total phosphates*

For its determination, sample was added to a test tube, along with 3 drops of solution A and 2 drops of solution B, manufactured by Indùstria Quimica Mascia Ltda (IQM), followed by shaking the tube. The color obtained by shaking was compared to the color chart for phosphate determination provided by IQM.

- *Total hardness*

For hardness, the sample was added to an erlenmeyer flask along with the buffer solution and the eriochrome black T indicator, and the hardness value was obtained by titration with the EDTA solution. The turning point for hardness was observed when the solution changed to a blue hue.

- *Sulphites*

To determine it, the sample, EDTA and sulfuric acid were added to a flask, followed by titration with potassium iodide-iodate until the solution turned blue. From the amount spent on the titration, it was possible to find out the amount of sulphite in the water.

- *Iron*

To determine the amount of iron in water, the sample was weighed in an erlenmeyer flask, followed by the addition of a 1:1 solution of phosphoric acid + sulphuric acid. Based on the titration method, the solution was titrated with potassium dichromate until it turned an emerald green color. From the amount spent on the titration, together with the weighed amount of the sample, it was possible to obtain the amount spent on iron.

- *Organic Matter*

The titration method was used to analyze organic matter. In this way, sample and sulphuric acid were added to an erlenmeyer flask and the solution was brought to the boil. When the solution began to boil, potassium permanganate was added, timed for 10 minutes, and then oxalic acid was added; after this, it was possible to titrate the solution with potassium permanganate. To determine this, the solution had to turn pink, so the amount spent on the titration was noted to determine the organic matter in the water.

It is worth noting that not all of the analyses carried out during the brewing process were carried out due to the length of the internship, which would have required more time.

5 FINAL CONSIDERATIONS

My internship at a brewing company was extremely important for my academic training, as during the internship I had the opportunity to correlate the theoretical lessons learned during my degree with the practical work carried out during the internship. I was also able to get involved in the routine of the laboratories, learning about the different types of beer and how their production works down to the smallest detail.

During my internship, I realized the importance of quality control within an industry, from the raw materials to the final product, always striving to produce excellent quality products.

Overall, it can be said that the internship period was enriching, as it provided more direct contact with the industry environment, as well as the responsibilities and duties that a quality control analyst has in brewing beer.

BIBLIOGRAPHICAL REFERENCES

AMBEV. Beers. Production. Available at: <http://www.ambev.com.br/produtos/cervejas>. Accessed on: October 11, 2018.

AQUARONE, E., et al. **Foods and beverages produced by fermentation. 4.ed.** Sâo Paulo : Edgard Blücher, 1993, 243p.

AQUARONE, E. **Biotecnologia industrial. 4. ed.** Sâo Paulo: Edgard Blücher, 2001, 523p.

AQUARONE, E.; ALMEIDA LIMA, U.; BORZANI, W. **Foods and beverages produced by fermentation.** Sâo Paulo: Edgard Blücher, 1983. 227 p.

BARBOSA, Thiago Muratori. Development of craft beer with yellow passion fruit pulp (Passiflora Edulis F. Flavicarpa Deg) and evaluation of the immobilization of Saccharomyces Cerevisiae cells in the alcoholic fermentation process. 2016. 55 f., il. Final course work (Bachelor's degree in Pharmacy), **University of Brasilia**, Brasilia, 2016.

BOGDA, P., KORDIALIK-BOGACKA, E. Alternatives to malt in brewing. **Trends in Food Science and Technology,** 65 (2017), pp. 1-9

BRAZIL. Ministry of Agriculture, Livestock and Supply. Decree No. 6.871, of June 4, 2009. Regulates Law No. 8.918, of 14 July 1994, which provides for the standardization, classification, registration, inspection, production and inspection of beverages. Official Gazette of the Federative Republic of Brazil, Brasilia, DF, June 5, 2009. Section 1. Available at: Accessed on: October 11, 2018.

BRIGGS, D. E. **Cereals and derived products.** In: DENDY, D. A. V.;

DOBRASZCZYK, B. J. Ed. Zaragoza. Acribia S.A 2001

BRODERICK, H. M.; CANALES, A. M.; COORS, J. H., et al. **El cervecero en la practica: un manual para la industria cervecera. 2.ed.** Peru: Associacón de Maestros Cerveceros de las Américas, 1977. 550p.

CARVALHO, G. B. M. de; BENTO, C. V.; SILVA, J. B. de A. **Fundamental biotechnological elements in the brewing process: Part 1 - Yeasts.** Revista Analytica, n.25, p. 36-42, 2006.

COOPER, C. M., EVANS, D. E., YOUSIF, A., METZ, N., KOUTOULIS, A. Comparision of the impact on the performance of small-scale mashing with different proportions of unmalted barley, Ondea Pro, malt and rice. **Journal of the Institute of Brewing** 122 (2) (2016), pp. 218-227

GREJO, Marcelo Matiolli. Determining the alcohol content of pilsen beer. 2014. 59 f. Final paper for the Industrial Chemistry course, **Fundaçâo Educacional do Municipio de Assis,** Assis, 2014.

INMETRO. Ministry of Development, Industry and Foreign Trade National Institute of Metrology, Standardization and Industrial Quality - INMETRO. INMETRO Ordinance No. 254 of September 18, 2009.

KUNZE, W. **Technology brewing and malting. Berlin: Research and teaching institute of brewing VLB,** 1996.

KUNZ, T., MULLER, C., MATO-GONZALES, D., METHNER, F. J. The influence of unmalted barley on the: Oxidative stability of wort and beer. **Journal of the Institute of Brewing, 118** (1) (2012), pp. 32-39.

LAUX, S. **Brewing. 2 ed.** 1997.

MADRID, Antonio. **Food industry manual: beer production process.** Sao Paulo, 1996, p. 285-313.

MARCONE, M. F., WANG, S., ALBABISH, W., SOMNARAIN, D., HILL, A. Diverse food-based applications of nuclear magnetic resonance (NMR) technology. **Food Research International,** 51 (2) (2013), pp. 729-737.

REINOLD, M. R. **Practical Brewery Manual. 1.ed.** Sao Paulo : Aden, 1997.213p.

REINOLD, M. R. **Water: the basis for good beer.** 2011.

SILVA, G. A., AUGUSTO, F., POPPI, R. J. Exploratory analysis of the volatile profile of beers by HS-SPME-GC. **Food Chemistry,** 111(4)(2008), pp. 1057-1063

TELES, J. A. **Study of the production of concentrated hopped wort from concentrated malt extract, high maltose syrup and hops.** State University of Campinas, Faculty of Food Engineering - FEA. Master's dissertation. Campinas, 2007.

VARNAM, A. H., SUTHERLAND, J. P. **Beverages: Technology, Chemistry and Microbiology.** Ed Zaragoz: Acribia S.A. 1997

VENTURINI FILHO, W. G. **Tecnologia de Cerveja.** Jaboticabal: Funep. 2000. 83 p.

VENTURINI FILHO, W. G. **Tecnologia de bebidas: matéria-prima, processamento, BFP/APPCC, legislaçâo, mercado.** Sao Paulo: Edgard Bliicher, 2005.

VENTURINI FILHO, W. G. **Bebidas alcoólicas v.1.** Sao Paulo: Edgar Blucher,

2010

ZSCHOERPER, Otto Paulo. **Brewer and maltster course booklet - AMBEV.** Porto Alegre: AMBEV, 2009. 71 p.

Printed by Books on Demand GmbH, Norderstedt / Germany